THE POETRY OF OGANESSON

The Poetry of Oganesson

Walter the Educator

Silent King Books

SILENT KING BOOKS

SKB

Copyright © 2024 by Walter the Educator

All rights reserved. No part of this book may be reproduced in any manner whatsoever without written permission except in the case of brief quotations embodied in critical articles and reviews.

First Printing, 2024

Disclaimer
This book is a literary work; poems are not about specific persons, locations, situations, and/or circumstances unless mentioned in a historical context. This book is for entertainment and informational purposes only. The author and publisher offer this information without warranties expressed or implied. No matter the grounds, neither the author nor the publisher will be accountable for any losses, injuries, or other damages caused by the reader's use of this book. The use of this book acknowledges an understanding and acceptance of this disclaimer.

"Earning a degree in chemistry changed my life!"
- Walter the Educator

dedicated to all the chemistry lovers, like myself, across the world

OGANESSON

Oganesson emerges, a marvel of mystery,

OGANESSON

Beyond the grasp of earthly bounds,

OGANESSON

In the realm where the rarest are found.

OGANESSON

In the nucleus's secretive embrace,

OGANESSON

Protons and neutrons interlace,

OGANESSON

A fleeting existence, transient and rare,

OGANESSON

Yet in its essence, an enigmatic flair.

OGANESSON

Born from stellar explosions profound,

OGANESSON

In the cosmic crucible, where elements abound,

OGANESSON

Oganesson emerges, a fleeting sprite,

OGANESSON

In the tapestry of the universe's might.

OGANESSON

With atomic number 118, it claims its throne,

OGANESSON

A noble gas, in isolation, it's known,

OGANESSON

Its electrons dance in quantum trance,

OGANESSON

A spectacle of the elemental dance.

OGANESSON

In laboratories, with meticulous care,

OGANESSON

Scientists coax it from the thin air,

OGANESSON

For a fleeting moment, it graces the stage,

OGANESSON

In experiments, its properties engage.

OGANESSON

A superheavy element, with mass extreme,

OGANESSON

In the periodic table, it reigns supreme,

OGANESSON

Yet its stability remains in question,

OGANESSON

A puzzle that defies comprehension.

OGANESSON

Oganesson, with its atomic allure,

OGANESSON

Captivates minds, seeking to explore,

OGANESSON

Its properties, its behavior unknown,

OGANESSON

In the realm where the rarest are sown.

OGANESSON

In the depths of the cosmos, it was conceived,

OGANESSON

In the crucible of stars, it achieved,

OGANESSON

A place in the pantheon of elements grand,

OGANESSON

A testament to nature's wondrous hand.

OGANESSON

Its name honors Yuri Oganessian's quest,

OGANESSON

For understanding elements, he did invest,

OGANESSON

In the pursuit of knowledge, he led the way,

OGANESSON

To uncover secrets hidden in the astral array.

OGANESSON

Oganesson, with your atomic grace,

OGANESSON

In the realm of elements, you find your place,

OGANESSON

A fleeting glimpse of cosmic design,

OGANESSON

In the endless expanse, where mysteries entwine.

OGANESSON

So let us marvel at this elemental sprite,

OGANESSON

As it dances through the fabric of the night,

OGANESSON

A testament to the universe's art,

OGANESSON

In the symphony of elements, playing its part.

OGANESSON

ABOUT THE CREATOR

Walter the Educator is one of the pseudonyms for Walter Anderson. Formally educated in Chemistry, Business, and Education, he is an educator, an author, a diverse entrepreneur, and he is the son of a disabled war veteran. "Walter the Educator" shares his time between educating and creating. He holds interests and owns several creative projects that entertain, enlighten, enhance, and educate, hoping to inspire and motivate you.

Follow, find new works, and stay up to date with
Walter the Educator™
at WaltertheEducator.com

www.ingramcontent.com/pod-product-compliance
Lightning Source LLC
LaVergne TN
LVHW012050070526
838201LV00082B/3884